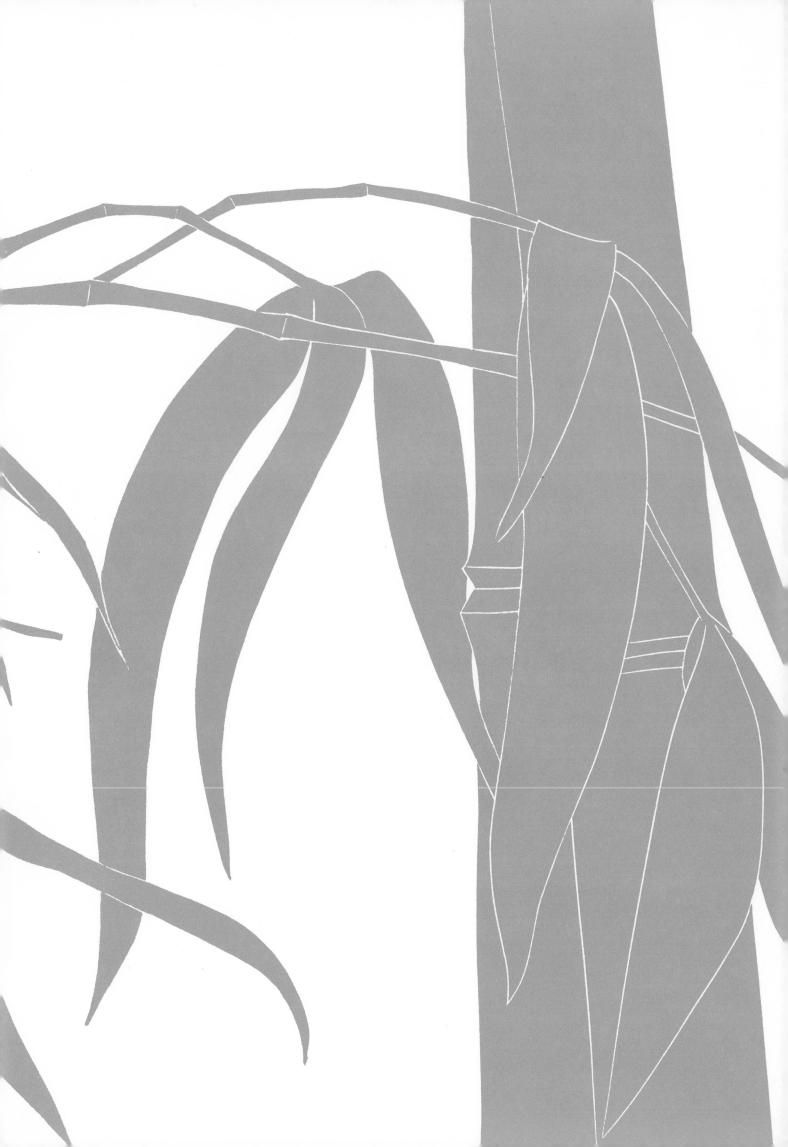

BOTANICAL PRINTS

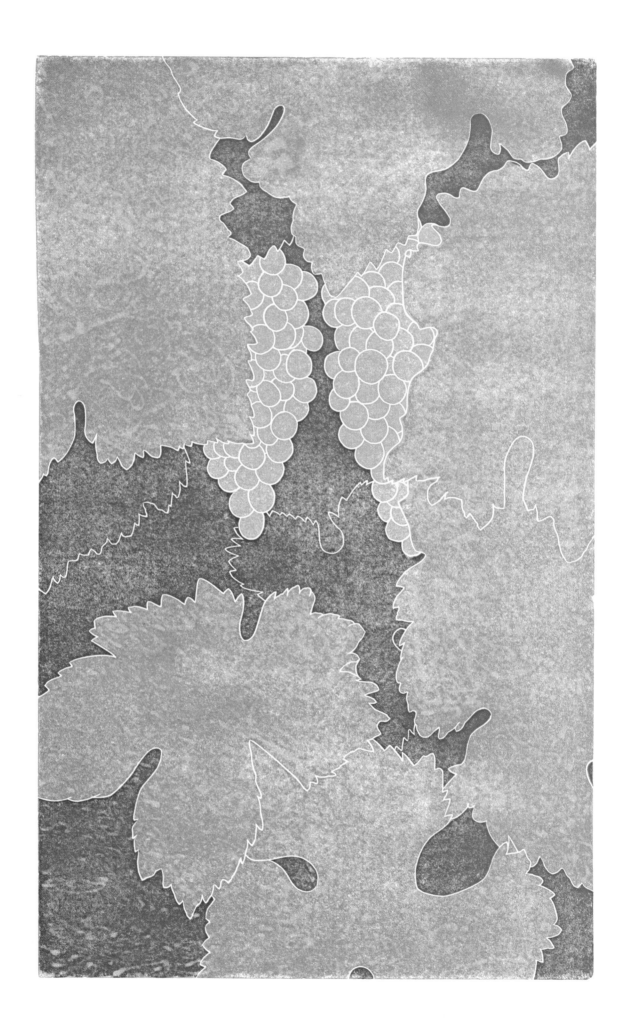

HENRY EVANS

BOTANICAL PRINTS

WITH EXCERPTS FROM THE ARTIST'S NOTEBOOKS

Foreword by
Wilfrid Blunt

W. H. Freeman and Company
San Francisco

ENDPAPERS Bamboo, species undetermined. Drawn at the Nikitsky Gardens near the Crimean city of Yalta, Ukrainian S.S.R. Printed in 1973 on Vit Bloma paper; 200 copies. Size of block: 25 cm x 40.1 cm.

FRONTISPIECE Traminer Grapes, a horticultural variety of *Vitis vinifera*. Drawn in the Napa Valley of California in 1974. Printed in 1975 on Basingwerk paper; 112 copies. Size of block: 40.7 cm x 25.3 cm.

Library of Congress Cataloging in Publication Data

Evans, Henry Herman, 1918–
   Botanical prints with excerpts from the artist's notebooks.

   1. Evans, Henry Herman, 1918–   2. Botanical
illustration. I. Title.
NE1336.E93.A43   769'.92'4   77-24323
ISBN 0-7167-0192-8

Printed in the United States of America

9 8 7 6 5 4 3 2

For Marsha, with love

FOREWORD

Most people, when they think of botanical illustration, visualize highly finished watercolor drawings of plants, and the very idea of using linocuts for the purpose may well seem strange; for it is a medium associated—in my country, at all events— with the immature artistic efforts of schoolchildren.

But how wrong can one be! Henry Evans, using this simple (though far from easy) technique, has produced portraits of flowers that can stand comparison with any that are being produced today in any medium whatever. The simplification that linocutting necessitates results, in the hands of such a master, in that strength that is so often lacking in the polished and labored watercolor.

Henry possesses the three qualities that are called for here (and, indeed, in all botanical illustration): sound botanical knowledge, an impeccable sense of design, and a complete sureness of touch. He is a master craftsman in a craft that permits of no fumbling and little correction. He also possesses a fourth and no less valuable asset: a wife who has learned, in no time at all, to print his linocuts to perfection.

It was about fifteen years ago that I first saw some of Henry's linocuts, and I was immediately captivated by them. But he has not been content to rest on his laurels, and year by year he has gone from strength to strength. His industry, too, is prodigious. With absolute confidence, therefore, I wish this book the great success it undoubtedly deserves, and I hope that it may prove to be the forerunner of many more.

Wilfrid Blunt

Compton, Surrey
1976

LIST OF SUBJECTS

BOTANICAL PRINTS

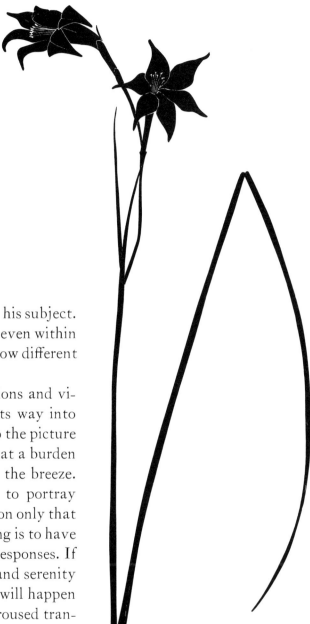

A special feeling comes when an artist perceives his subject. Each artist relates to his subject matter differently; even within this narrow field, each artist is still himself, is somehow different from any other.

The feelings and values we bring to our perceptions and visions are the result of everything that has made its way into our consciousness. But to bring everything we are to the picture would be to create a jumble. It would be far too great a burden to impose on some delicate little flower swaying in the breeze.

What we hope to do when we are attempting to portray beauty is to let through into the moment of perception only that part of us that is essentially beautiful. The ideal thing is to have the subject speak to us and be the conjuror of our responses. If the image is beautiful and serene, then the beauty and serenity that is in us should be ready to respond to it. This will happen only if we are able to bring ourselves to a state of aroused tranquility. Such a coexistence of opposites is not impossible, but it does demand discipline.

ABOVE Homeria, probably *Homeria elegans*. A South African plant found and drawn at the University of California Botanic Garden, Berkeley. Printed in 1974 on Basingwerk paper; 250 copies. Size of block: 41 cm x 20.1 cm.

Conceding this, imagine coming upon a perfect wild rose in quiet summer woods. The flower does not evoke the past, only the infinite possibilities of the present. The sun is straight above and the rose's tiny thorns point in every direction. As you change your perspective, the color of the petals seems to change —soft pink, dark pink, red. The leaves, arranged in formal groups like tiny motionless fish, are dark above and light below, with minute serrations along the edges. There are so many irregular curves in the stem. Somewhere nearby a deer is walking, crushing dry leaves and breaking dry twigs.

The pencil is now touching the paper. The first mark on the paper is like the last touch of your toes on the diving board. So many decisions have already been made by now that the drawing almost seems like the end of the process, not the beginning. Deciding where to stand when you draw is probably the most critical. Then you must decide how much of the plant will be in the drawing and which parts within the chosen area to portray. You may leave things out when you draw from life and often improve the composition by doing so, but it is folly to try to add what is not there.

The pencil moves faster because the decisions have been made —only a few lines are needed here and one or two there. The deer is still stumbling around somewhere in the woods in back of you. Wasps have come to investigate your pack. Some hikers with a barking dog are coming down the trail. The spell is broken, and everything is again as it was before you saw the rose.

ABOVE   Eucalyptus, *Eucalyptus globulus*. Drawn from a specimen collected on the west side of Angel Island in San Francisco Bay. Printed in 1973 on Vit Bloma paper; 200 copies. Size of block: 40.3 cm x 12.5 cm.

OPPOSITE   Wild Rose, *Rosa virginiana*. Drawn at Castine, Maine. Printed in 1971 on Hodamura paper; 125 copies. Size of block: 25 cm x 21 cm.

Stages of Drawing

When I have made a drawing on paper in the field, I bring it back to the studio where, when the right time arrives, it will serve as the basis for a new drawing made directly on the linoleum block. (The drawing that is made on the linoleum *can* be made directly from the subject, but I have found that during the making of the second drawing I can definitely improve upon the first.) This second drawing has to be a distillation of the original experience—with a greater awareness, perhaps, of the technical problems that must be solved in making the print. I must use a fairly soft pencil to avoid bruising the surface of the block. I know from experience, however, that this second drawing is also an intermediate stage, that the determination of the final form will be made in the cutting. It's still possible to consider making changes that are truer to nature, more pleasing to the eye, or closer to the limitations of the medium. Every art form has its limits of detail. Not that drawing every hair in exquisite three-dimensional clarity makes the image any more believable, but there must be enough characteristic detail to communicate the essence of the subject.

While making the second drawing on the block, I have a very welcome opportunity to reevaluate my original attempt, to be self-critical from the vantage point of some elapsed time. (It's always easier to face one's own errors later, rather than at the time they are committed. In fact, at the moment of commission, one usually cannot see them.) So a week or two after I have made the original drawing, I can sit down with the block, and the drawing, and a pencil (and an eraser) and make a drawing on the block that is an improvement over the original.

Above   Teasel, *Dipsacus fullonum.* Drawn near Cronkhite Beach, Marin County, California. Printed in 1973 on Basingwerk paper; 260 copies. Size of block: 40.5 cm x 22.5 cm.

Opposite   Sahuaro Cactus, *Carnegiea gigantea.* Drawn in the desert near Tucson, Arizona. Printed in 1968 on Hodamura paper; 96 copies. Size of block: 31.7 cm x 21.5 cm.

Cutting the Block

The whole cutting process is basically different from drawing, although it's true that the same kind of intense concentration must be employed in both processes. The difference is that the hand with the pencil makes a gesture that is *not* final. A line can be erased and redrawn many times, if need be. But once a piece is cut out of the block, it can't be put back.

When the cutting edge of the blade comes down to touch the surface of the block, there are myriad considerations fighting for first place in my mind. Whether to make a curve larger or smaller, a stem fatter or thinner, a leaf more or less curved—none of these is a simple decision. Even though I may not be conscious of it, the battle is still being fought with each gesture during the critical stages of the cutting. Once the form has been delineated, cutting becomes more a matter of concentration and technique and less a matter of choice and taste. The first stage of the cutting is done with the finest tool, and once it is done the form can be altered very little.

Tool Sharpening

Tool sharpening is critical: good work cannot come from dull tools. The smaller the tool, the harder it is to learn to sharpen it—but this is a lesson you may not avoid. You must learn to use the Carborundum, India, and Arkansas stones and work until the tiny gouges are razor sharp. A dull tool has a mind of its own. You cannot control it. The sharp tool does as it's told. That's an oversimplification, of course, but it's the beginning point in printing from blocks. It took me several years to gain an appreciation of tool sharpening and to begin to master the skills, but I found I couldn't refine my efforts with dull tools. I was desperate to get closer to expressing my own visual perceptions, so somehow, through books, the advice of friends, and my own intuition, I figured it out in a way that works for me.

Above Spider Chrysanthemum, probably a form of *Chrysanthemum morifolium* (or *Chrysanthemum hortorum*). Drawn in Fremont, California, from a specimen grown by Mike de Salles. Printed in 1973 on Basingwerk paper; 500 copies. Size of block: 40.5 cm x 23.5 cm.

Opposite Black-eyed Susan, *Rudbeckia hirta*. Drawn at Saratoga Springs, New York. Printed in 1974 on Basingwerk paper; 230 copies. Size of block: 40.6 cm x 25.2 cm.

Feelings and Directions

Only certain feelings that an artist brings to his work can be effectively expressed with a given subject. There is no way of putting the burden of one's social consternation on the shoulders of one little wildflower, at least not for me. So the print I make cannot express my total person. Making the print is a very selective process, a conscious effort to show the beauty and elegance of plant forms as an expression of my admiration for the wonders of life on this planet.

Limits of the Medium

Having arrived at a point of conviction that the printmaking I do is a selective process, I can then proceed to push to the limit the medium I have chosen. In the early years of my still continuing apprenticeship, I heard many warnings about what you could and couldn't do with linoleum—it will break, it will wear out, or you can't cut it this way or that way. All of the warnings proved to be wrong, not because I am uncommonly smart or exquisitely endowed, but because I may have been a little more stubborn than most people and because I was so nauseated by the idea of failure. Some people are born problem solvers, and, for better or worse, I am one of those people. If something I try to do doesn't work, I get a little sore at everything and hate to give up.

I experienced the same awful struggle with learning how to mix colors. Most printers do not mix their own inks. They order ready-mixed colors from an ink manufacturer's catalog. But the ink as it comes straight from the can is never quite right, and I soon found out that learning to mix the right color can be very costly. It took a lot of persistence to come to terms with the rainbow, which seemed to me so strange and inexplicably magical. Who in the world could really make it clear why yellow and blue together make green? Or why putting one or the other underneath or on top would result in two totally different colors?

Above   Wheat, *Triticum* hybrid. Drawn in the studio from a dry specimen collected in the Sacramento Valley of California by Duane Onomiya. Printed in 1974 on Basingwerk paper; 250 copies. Size of block: 41 cm x 24.2 cm.

Opposite   Bitterroot, *Lewisia rediviva*. Drawn at Berkeley, California. Printed in 1968 on Hodamura paper; 72 copies. Size of block: 20 cm x 25 cm.

[8]

Color Mixing

My original idea of color relationships and color mixing dates back to grammar school. The teacher had tacked up a color wheel which she pointed to with an old yardstick as she tried to convey to us ideas about primary and secondary colors, and complements and opposites, and shades and hues. All of this was exceedingly remote stuff to a nine- or ten-year-old boy who was not too eager to make pictures. The urge to make pictures came later, and with it came a consuming curiosity about color. I had a vague recollection of that day long before—my first exposure to ideas about color as a school child—and I went to consult the textbooks, full of hope. When I tried to put my textbook knowledge into practice, I ended up with great sloppy batches of ink that did not even remotely resemble what I wanted. So I had to learn about the practical part of mixing color as well as the visual and theoretical parts. It took a lot of trial and error, and a lot of wasted ink, before I began to get the hang of it.

Color Overlays and Color Relationships

I also learned very early in the work that blocks printed over each other produce many surprises. Because you are working by eye and not by formula, and because you cannot apply the ink to the block in precisely equal amounts for each impression, there will be very slight color variations from print to print when there are overlays of color. With three or more blocks, it is possible to make an edition of 100 prints in which *all* of them are perceptibly different from one another.

There is a way to keep closer control of the colors: cut the blocks with no overlapping of colors at all, so that each color area is separate from the others. There still may be some slight variation in color from print to print because of the varying *quantities* of ink you get onto the paper, but this variation is not magnified by the overlapping of the blocks.

ABOVE   Bamboo, *Bambusa multiplex* 'Goddess.' Drawn from a specimen that I grew in our apartment in San Francisco. Printed in 1976 on Basingwerk paper; 190 copies. Size of block: 31 cm x 23.2 cm.

OPPOSITE   Chenin Blanc Grapes, a horticultural variety of *Vitis vinifera*. Drawn in the Napa Valley of California in the Fall of 1974. Printed in 1975 on Basingwerk paper; 145 copies. Size of block: 38.5 cm x 25.2 cm.

PAGES 12 AND 13   Grass-Iris (or Blue-eyed-Grass), *Sisyrinchium bellum*. Drawn on Angel Island in San Francisco Bay. Printed in 1973 on Basingwerk paper; 150 copies. Size of block: 20 cm x 35.7 cm.

Choosing Paper

In the early stages of my work as a printmaker, I was deeply impressed by the way different papers would receive the same print. I would use the same ink and would apply the same amount of pressure to the block, but the results would be quite different. Let me qualify this: to the casual viewer with a relatively untrained eye, the differences might not seem great, but to the person working closely with the print these differences never fail to be impressive.

A paper's color, texture, and receptivity to ink all affect its suitability for printmaking. Colored papers invariably fade with exposure to light. I learned very early that the lovely hand-made Japanese papers that come in various colors, such as the Toyamas, fade very quickly and unevenly. Some people accepted this as added charm—even doted on it—but it never pleased me very much and it disappointed, even angered, some of my patrons. In all my printing on various colored papers, I never found one that didn't fade or streak or at least lose some of its original freshness.

The search for an ideal sheet of paper with which to work is a long and difficult task. I spent years trying to get the very best papers that I could afford. Even in the early years of my work, all of the good papers were expensive. Since then, many of the outstanding mills have ceased to exist, so the range of choices today is much smaller, and the good papers are extremely costly. I have settled on Basingwerk as a day-to-day sheet. It is made by Grosvenor Chater in England; it contains esparto fiber, which comes from a grass grown in North Africa, and it is eminently suitable for fine printing. It takes the ink embarrassingly well. In fact, *everything* prints, down to the tiniest details in the surface of the block, so preparing the block for printing may take considerable time and effort.

ABOVE Grass, species undetermined. Found growing about six feet above high water, in a handful of dirt on a ledge of rock on the coast a few miles north of Stewarts Point, Sonoma County, California. Printed in 1976 on Basingwerk paper; 240 copies. Size of block: 38.8 cm x 25.5 cm.

OPPOSITE Blue Larkspur, *Delphinium patens*. Drawn at Stanford University's Jasper Ridge Biological Reserve, San Mateo County, California. Printed in 1976 on Basingwerk paper; 240 copies. Size of block: 40.5 cm x 17 cm.

Because I have used this Basingwerk sheet for some years now, I am well acquainted with its characteristics. This is an enormous advantage. I would advise any beginning printer or printmaker to try to find a sheet of paper that works well for him or her, and is financially within reach, and then to stay with it. Each paper has a personality and a set of behavioral characteristics all its own, and it takes time to learn all of its little quirks.

One of the main reasons for using only one kind of paper is that you will, in any event, have plenty of other variables to deal with. In printmaking, there is no end to difficult problems. The best approach is to try to establish things in your work cycle that are *constant*. New things come up every day—new problems that you didn't expect, didn't know about, had never heard of—so establishing constants is to your advantage.

Choosing Inks and Using Them

It also makes sense to find a brand of ink you can work with comfortably and then to stick with it. Inks that use linseed oil as a vehicle do vary, but most ink companies make what they call a "series of colors." If the ink is made well, all the colors in the series are of much the same consistency and will mix well with each other. (The key to this is that the oil in each of the inks has been cooked to the same consistency. It is also important that the pigments be finely and evenly ground.) The ink maker will usually also offer a thinner and a drier that have been formulated especially for use with that series of inks.

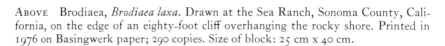

ABOVE Brodiaea, *Brodiaea laxa*. Drawn at the Sea Ranch, Sonoma County, California, on the edge of an eighty-foot cliff overhanging the rocky shore. Printed in 1976 on Basingwerk paper; 290 copies. Size of block: 25 cm x 40 cm.

OPPOSITE Iris, *Iris* hybrid. Drawn in the studio in San Francisco. Printed in 1974 on Basingwerk paper; 130 copies. Size of block: 40.2 cm x 24.9 cm.

Early in my work, I discovered, after a number of ludicrous and costly experiences, that you need a lot of good white ink. When you mix colors, the best way to start is with white, perhaps with a tenth or a fifth of the whole quantity of ink you think you might need. The quantity is not easy to judge at first, but experience will sharpen your judgment. It's a lot easier to mix enough at first than to try to mix a second batch exactly the same as the first. But you can easily become the sorcerer's apprentice if you need a light hue of a color and you start with the color and add white to it. Always start with white and add the color to it gradually.

Early on, I began trying to avoid the waste of ink. Ink doesn't really dry—it sets up or hardens like paint. Ink left out on a piece of paper, in a plastic cup, or in a tin can will form a skin. Storing a can of ink that has been opened and partly used always involves some loss. Each time you take ink from that container, you have to remove some of the skin to get to the usable part underneath. The bit of skin is waste, and afterwards a new skin begins to form in its place on top of the unused remainder of the ink in the can, and so on until the container is empty. Because exposure to the air is what causes the surface to set up, some printers fill their ink containers with water during storage. They throw the water away before trying to remove any ink from the can and replace it when they have taken what they want. It helps, and some printers think it's worth the trouble. But generally, there is some unavoidable waste of ink that you have to learn to live with. Ink, after all, constitutes only a minute part of the cost of making a print.

ABOVE Salal, *Gaultheria shallon*. Drawn in the studio from a specimen grown by Gunder Hefta. Printed in 1973 on Basingwerk paper; 58 copies. Size of block: 17.5 cm x 38.3 cm.

OPPOSITE Eggplant, *Solanum melongena*. Drawn at Ukiah, California. Printed in 1975 on Basingwerk paper; 120 copies. Size of block: 39 cm x 25 cm.

Paper Quality

Paper waste is quite another matter. If you are trying to earn your living by making prints, you can't help resenting the loss of every sheet that is spoiled. Yet you will risk thwarting your own efforts if you try too hard to be economical: if you pull proofs on inferior (cheaper) paper, they won't really tell you what you want to know. The most helpful and informative proofs are pulled on the same paper on which you intend to print your edition.

You can avoid some of the agony of wasted paper by sorting through each ream or package of paper before the printing begins. In this way the sheets with the most repulsive (eye-catching) flaws can be set aside for trial and experimental proofs. You already know that some of the early prints you pull are going to be unsignable and unusable, so this is a far less painful way to use the imperfect sheets.

Over the years I've had a lot of experience with irregularities in paper. When I was just getting started, and sales were not nearly so good, I signed every impression my conscience would allow me to sign. In those days, I was using a lot of handmade Japanese papers in which the irregularities (I'm not at all sure they were flaws) were very numerous—even in some of the Hodamura whites, where the idea of a plain white surface was probably considered but seldom achieved. There were some unbelievable things made right into the paper, which symbolized captured moments in time if nothing else—strands of the papermaker's hair, complete insects, unlikely bits and pieces of all descriptions—but each one, quaint and curious as it was, actually spoiled a piece of paper.

ABOVE Dry Pods, probably fruits of the Iceland Poppy, *Papaver nudicale*. Drawn in the studio from specimens brought in by David Crossman. Printed in 1975 on Basingwerk paper; 125 copies. Size of block: 40.5 cm x 25.2 cm.

OPPOSITE Anemone, probably *Anemone coronaria*. Drawn in France in 1973. Printed in 1974 on Basingwerk paper; 140 copies. Size of block: 40.5 cm x 24.8 cm.

Minor impurities in the pulp that managed to get made into a sheet were sometimes close enough to the edge that they could be hidden by the mat, and some would be sufficiently covered and camouflaged by the ink to be of no consequence. Still, many people have very strong ideas about the immaculate, and it's best not to argue with them.

Choosing Subjects

Within any given range of subject matter, the artist can usually find, as I have, an enormous assortment of individual subjects to draw. For reasons not too easy to explain, some subjects are more appealing to the public than others. Equally mysteriously, when an artist is drawn to a subject, and perhaps puts more of himself into it, the public tends to respond more deeply, more enthusiastically. Some sort of mutual-encouragement syndrome seems to develop. When you manage to convey your all-too-human feeling in a work of art, certain people who see it will vicariously experience some of the intensity of your involvement and will become absorbed by, and possessive of, what you have done. When this is combined with a very familiar or sentimentally loaded subject, the response is even stronger and the desire to possess the picture seems to increase.

I know now, through years of experience, that certain plants make far more meaningful pictures than others. Some subjects are practically assured of success before the work is begun. The mutual-encouragement syndrome takes hold quite imperceptibly at first. The ego thrives on compliments and kind words, especially regarding something you liked very much in the first place. It is particularly nice to have one's own feelings, sentiments, and enthusiasms approved of, even praised. So, not long after, you find yourself drawing again what brought you pleasure and success before. Thus one might unwittingly change one's goals as an artist—which is not necessarily a bad thing.

ABOVE   Freesia, *Freesia hybrida*. Drawn at the Bristol Hotel in Paris. Printed in 1974 on Basingwerk paper; 83 copies. Size of block: 40.3 cm x 21.7 cm.

OPPOSITE   Tulips, *Tulipa gesneriana*. Drawn at the Crillon Hotel in Paris. Printed in 1975 on Basingwerk paper; 105 copies. Size of block: 40.5 cm x 24.5 cm.

The urge to return to the same subject can stem from a sense of not having done as good a job as perhaps one might have, had all the circumstances been better. But there is also the possibility that you have found a comfortable rut, or worse, that you have ceased to be creative. Among my own friends and acquaintances are artists who have saved all their blocks to reprint as the need arises. They do not number the prints so, as long as the printing surface is not worn or injured, the printing can continue indefinitely. The hazard here, and it is a great one, is that the artist will grow perfectly content to reprint his old blocks indefinitely, and so will make no new ones.

Collectors

A very important aspect of the artist's relating to the public is that individuals and institutions may put together collections of his work. The individuals who collect may or may not consider themselves collectors. The institutions, including museums, libraries, and private corporations, are managed by individuals, so the artist is still dealing with people. But the artist can easily get edgy about all this. Collecting for aesthetic reasons is not at all as common as collecting for economic ones, and the sensation that someone has a purely financial interest in your work is a peculiar one. Oddly, the most critical event in an artist's life, as far as the economic value of his work is concerned, is his own death. People have said to my face: "When you die, boy, your stuff is going to be worth a lot!" I can hardly wait.

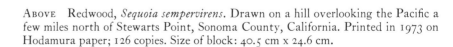

ABOVE   Redwood, *Sequoia sempervirens*. Drawn on a hill overlooking the Pacific a few miles north of Stewarts Point, Sonoma County, California. Printed in 1973 on Hodamura paper; 126 copies. Size of block: 40.5 cm x 24.6 cm.

OPPOSITE   Tiger Lily, a horticultural variety of *Lilium*. Drawing begun in Madrid and finished in Barcelona the following day. Printed in 1976 on Basingwerk paper; 245 copies. Size of block: 40.5 cm x 20.5 cm.

The Price of Art

I have on occasion felt a deep sense of envy when, walking through a market, I have come upon a skilled artisan displaying his wares, offering them for sale at prices even the most poorly paid people can afford. Such an artisan knows that his work has a use and a function, and that it can also be admired for itself alone. Does our society tend to preclude the possibility of this simple pleasure? I hope not. Thus far, by keeping snob appeal minimal and prices down, I have kept an enviable relationship with the people who buy my work. It's not expensive enough to be exclusive and it's not sensational enough to be disconcerting. I have been saved from becoming part of the fashionable "art world" of galleries, cocktail parties, and speculation. Earning a living by making pictures to sell for low prices takes a lot of time and simply doesn't leave many hours for toadying up to prospective patrons. When I sell a low-priced print, I only have to say "thank you" once and that's the end of it. I can get back to work, or carrying out the trash, or blowing bubbles, or whatever it is that I am doing.

Making Transfers

When a print must have more than one color, a number of alternative methods exist for printing the additional colors. Occasionally, if the relationship of the color areas makes it physically possible, the printer can cut the original block into two or more parts and print them separately. Where the colors are contiguous, and they generally are at one or more points, the cutting must be done with the thinnest blade and the finest point you can find, in order to avoid marring the joining edges of the design. After the first color is printed, the next block must be fitted very carefully into place before the first one is pulled off the base.

ABOVE Leather Fern, *Polypodium scouleri.* Drawn in the studio from a specimen grown by Gunder Hefta. Printed in 1973 on Vit Bloma paper; 190 copies. Size of block: 35.9 cm x 24.2 cm.

OPPOSITE Chrysanthemums and Chicory, *Chrysanthemum morifolium* and *Cichorium intybus.* The chrysanthemums were drawn in Italy and the chicory was added later from a specimen found at Suisun City, California. Printed in 1974 on Basingwerk paper; 135 copies. Size of block: 39.5 cm x 25 cm.

PAGES 28 AND 29 White Pine, *Pinus strobus.* Drawn at the United States National Arboretum, Washington, DC. Printed in 1971 on Hodamura paper; 110 copies. Size of block: 25.5 cm x 41.5 cm.

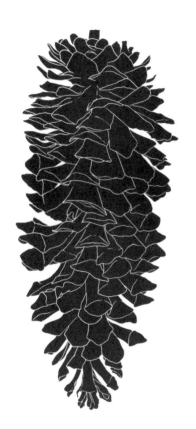

Usually, however, one or more transfer blocks must be made. On a hand press, this can be done easily and quickly: After cutting the original design on the first block (using your finest tool and removing none of the background), put the block into the press and ink it rather heavily with white ink. Then take an impression on smooth paper, and let the paper you have just printed stay in the press. Next, remove the block from the press and replace it with another block that is clean and new. Bring the printed sheet down on the fresh block and "print" again. If the pressure is adequate and the inking right, there will be a very clean transfer impression on the new block: the lines you cut into the original block will be unprinted on the transfer block and the rest of the block will be covered with white. Once the transfer block is dry, it can be cut for the printing of the second color.

For each color beyond the first that you want to add to your print, you must have another transfer block. Cutting the transfer blocks is a very delicate matter. Until you have acquired some skill, it is safer to make one or two extra transfer blocks: then if you make a mistake, you have another one on which to try again.

It was in trying to learn to cut two or more blocks to fit with each other exactly that I began to appreciate how difficult it really is to develop sufficient physical control of the cutting tool. I sometimes look at the work of other artists in galleries, museums, and friends' homes. Much of it seems intentionally casual, and I must confess I am sometimes a little envious of the happy carelessness of other artists. I am quite convinced that there is some sort of virtue in "good clean work," but I'm not sure I could tell you exactly what it is. The artist must listen to the voice inside him. Mine seems to abhor carelessness. The voice is always there, so I guess that's the way it has to be.

ABOVE Sugar Pine, *Pinus lambertiana*. Drawn in the studio from a specimen collected in California's Mother Lode country. Printed in 1973 on Vit Bloma paper; 202 copies. Size of block: 40.6 cm x 18.2 cm.

OPPOSITE Brodiaea, *Brodiaea pulchella*. Drawn in Berkeley, California. Printed in 1973 on Basingwerk paper; 220 copies. Size of block: 39.5 cm x 24.6 cm.

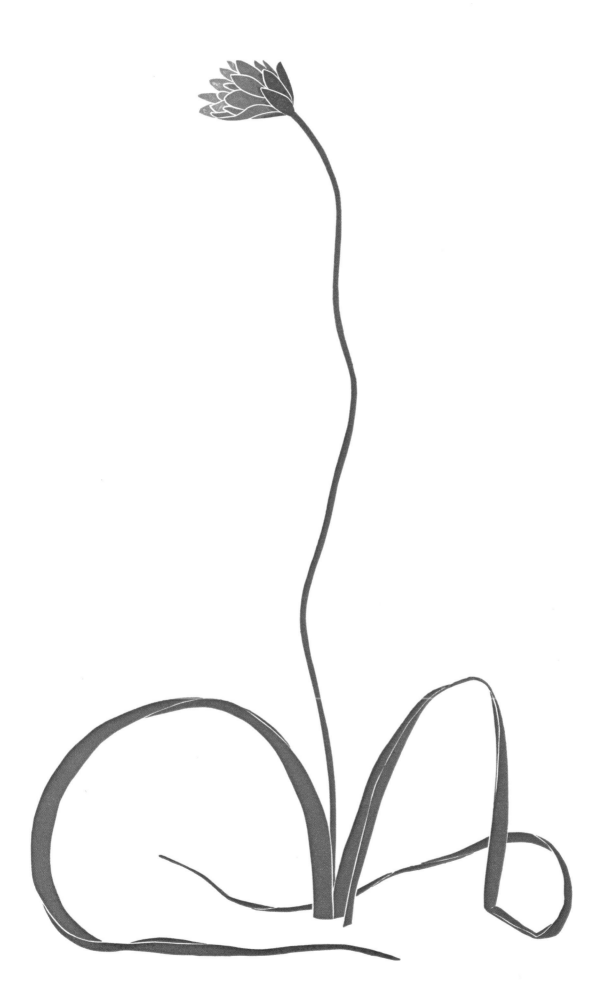

Superimposing Colors

Subtle shades and hues of color are so important to our recognition of some plants that a print of such a plant is successful only if it suggests them. Sometimes a simple abutment of solid colors will create the desired effect. If it does not, sometimes the problem can be solved by arranging to have blocks (or parts of blocks) overlap during printing. Printing with overlays of color always requires two or more blocks. If you use more than three or four blocks, the amount of time required for printing is considerable, especially if the edition is large.

Overlaying of colors is a very complicated and personal process. I make a transfer for each block beyond the first, and sometimes I make one or two beyond that just to be on the safe side. But before I begin, I always separate the color areas first in my own head so that I can visualize what it is I need to do in cutting the transfer blocks. The whole business of "color separation" in one's own head is tricky to say the least, and I would think that for many people it would be the most difficult aspect of printmaking.

As you begin the cutting, you may be a little nervous because you know that once you have cut the block you can't back up and start over. Each time you push the tool and gouge out a piece you have made a very definite statement about the ultimate shape and final effect of the print. You can start again with a new block, of course, but the more complex the project and the more work it demands the less likely it is that you will want to begin again. Of course, when the effect is unacceptably far from what you had visualized, you *must* start all over again.

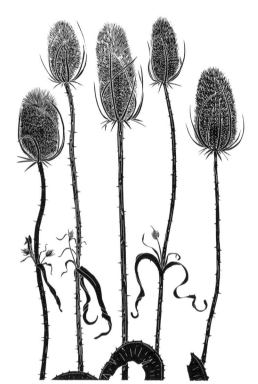

ABOVE Teasel, *Dipsacus fullonum.* Drawn in the studio from specimens collected near Hopland, Mendocino County, California. Printed in 1975 on Basingwerk paper; 300 copies. Size of block: 39.2 cm x 25 cm.

OPPOSITE Rose, *Rosa* hybrid. Drawn in London. Printed in 1974 on Basingwerk paper; 160 copies. Size of block: 40 cm x 24.7 cm.

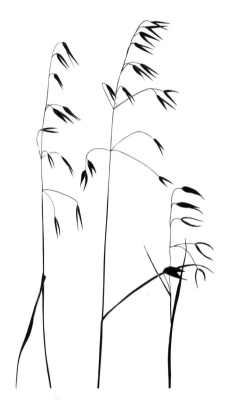

Composition, Motion, and Dynamics

When I compose a picture in my mind, I try to consider the viewer's perception of it. High on my list of considerations is the fact that, in our culture, because we read from left to right, the natural motion of the eye is from left to right. Consequently, if there is to be a strong lateral movement in the composition, I make it go from left to right. Pictures made by people of other cultures do have a different look. When you remember that some Oriental languages are written and read from top to bottom, you can see why so many Oriental pictures are vertical, even emphatically so. Similarly, Israeli art may owe much of its distinctive look to the fact that Hebrew reads from right to left.

The so-called dynamics of a picture must be related to natural eye motion, but that is only one element of understanding and comprehending the picture. Where the picture begins and ends is of critical importance. Simply stated, if too many strong elements are cut off at the edges of the composition, the viewer's eye cannot complete the composition and it tends to fall apart. The central forms and movements lead the viewer to *expect* certain things to be present, and if they are chopped off the picture is too hard to believe. A certain feeling of completeness or wholeness is necessary to make the picture be credible. Of course, mats and frames can sometimes strengthen the credibility of a composition.

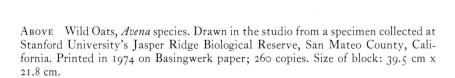

ABOVE   Wild Oats, *Avena* species. Drawn in the studio from a specimen collected at Stanford University's Jasper Ridge Biological Reserve, San Mateo County, California. Printed in 1974 on Basingwerk paper; 260 copies. Size of block: 39.5 cm x 21.8 cm.

OPPOSITE   Shooting Star, *Dodecatheon hendersonii*. Drawn near the bank of Robinson Creek, southwest of Ukiah, in Mendocino County, California. Printed in 1976 on Basingwerk paper; 230 copies. Size of block: 33.5 cm x 21 cm.

Familiarity

I think that familiarity with the subject can be very important to the viewer's impression of a picture. Visual xenophobia seems to be an almost universal condition. If a picture is strange, then most people are either repelled or simply uninterested. The more familiar and pleasant a picture is, the more people will like it. Frost this familiarity with sentiment and you have a winner. Familiarity and sentiment are often inextricably intertwined. Many artists cannot do this at all. Some (brush in cheek) can when necessity dictates. Some do it as a perfectly natural act, like breathing. Successful communication of ideas and feelings in art is fraught with many matters of taste and value. The immortal dignity and elegance of the Egyptian bas-relief is not necessarily *better* than the syrupy greeting card with cuddly bunnies: it's different. There are times when cuddly bunnies are more appropriate. And there are artists who can do them.

Drawing and Redrawing

Drawing and redrawing are the rudiments of a process that is very basic to making prints. Once I have selected the subject and decided upon the composition, I proceed to sketch on paper the elements of the form. While this sketch is being made, I make note of certain characteristic structural details. Sometimes I even draw parts of the sketch in much greater detail, or I draw a little close-up in the margin with more than the usual amount of detail clearly shown: the special ways that joints are formed in this particular plant, the peculiar ways that leaves are fastened onto the twigs, or the way that the flower petals curve.

ABOVE Calochortus, *Calochortus albus*. Drawn at the University of California Botanic Garden, Berkeley. Printed in 1973 on Basingwerk paper; 115 copies. Size of block: 39.5 cm x 25.2 cm.

OPPOSITE Peach, *Prunus persica*. Drawn at Modesto, California. Printed in 1970 on Hodamura paper; 80 copies. Size of block: 41 cm x 24 cm.

The redrawing from paper onto the block is the point at which basic shapes can be improved and changes can be made in the composition to accommodate more logically the limitations and peculiarities of the medium. Even the drawing on the linoleum is not the last chance you have to correct and improve according to your taste and vision, but it is your last chance to try to remember the actual form and stance of the plant back there in the woods.

Recalling Form and Color

The recollection of things past is for some people painfully difficult, if not impossible. Others seem to have better recall. Perhaps the ability to remember is an acquired (developed) skill. For me, the best sort of memory is, to transmute a phrase, tranquility tranquilly recalled. At the risk of turning poet's gold into base metal, it seems to me that the whole ambiance of the original experience is critical to what one can recall at a later moment. The less perfect the original circumstances, the more difficult the experience will be to remember. We repress the unpleasant memories as we hope for immortality for the good ones.

Remembering color is a stage beyond remembering form, just as dreaming in color is less common than dreaming in black and white. I think this probably has a physiological basis. Human beings are variously endowed with sight. Not only do we see actual colors differently, we also see intensities of color and color contrasts differently. For some people, color recall seems to be almost impossible. I think people who have an average or better-than-average color perception can *learn* to recall colors, but it is a slow process and it requires much practice in thinking about color in a critical, evaluative, comparative way. Trying to mix a color to match your recollection is very difficult at first, but, over a period of years, any person of average perceptions can learn to do it.

Above Grass, species undetermined. Drawn at the Sea Ranch, Sonoma County, California. Printed in 1976 on Basingwerk paper; 135 copies. Size of block: 40.3 cm x 21.2 cm.

Opposite Iceland Poppy, *Papaver nudicaule*. Drawn at Nice, France. Printed in 1975 on Basingwerk paper; 120 copies. Size of block: 40.5 cm x 25 cm.

Refinement

I can recall when I first saw Matisse's illustrations for the Limited Editions Club edition of James Joyce's *Ulysses*. I remember being intrigued, even then, with the idea that an artist could move in stages through a succession of his own drawings and redrawings toward an essential, perhaps quintessential, expression of an image or idea. As with many impressionist works, I now feel that, for better or worse, this shifts the responsibility to the viewer for providing all details and context. The viewer must conjure up the very fabric of the picture himself. Leaving aside the fairness of it—for to confuse art and ethic is irrelevant—I wonder how the artist decides where his work will leave off and the viewer's work will begin. I am certainly not opposed to delegating parts of the creative process, but whether or not it is necessary probably depends on how clearly the artist conceives the idea.

Like many artists, I redraw certain drawings that I have strong feelings about in hopes of adapting them to further use. The difference is that I do not try to discard detail in successive stages, but rather to adapt the detail to the next stage of the drawing.

When you are making prints from blocks you have cut, you always have to keep in mind the fact that in the printing process the image is subjected to a left-to-right reversal: what appears on the right side of the block itself will be on the left side of the print when the print is made on the paper. Because natural eye motion in our culture is from left to right, each time you redraw you have to consider what your changes will mean to the visual dynamics and movement in the picture.

ABOVE Daisy, *Chrysanthemum leucanthemum.* Drawn from a specimen purchased from my friend Albert Nalbandian at his flower stand on the corner of Stockton and Geary in San Francisco. Printed in 1974 on Basingwerk paper; 180 copies. Size of block: 40.5 cm x 21.2 cm.

OPPOSITE Shasta Daisy, a horticultural form of *Chrysanthemum maximum.* Drawn at home in San Francisco. Printed in 1973 on Basingwerk paper; 147 copies. Size of block: 40.3 cm x 17 cm.

I have found that I tend to draw slightly larger at each stage. I haven't figured out yet whether I am just being lazy and making it easier for myself to deal with fussy details, or whether I have imperfect hand-to-eye coordination, but when I redo a drawing a number of times it becomes larger than life size. There is some loss of detail as the progressive decisions are made about what is really essential, but it is minimal. Redrawing for me is a means of refinement rather than abstraction.

Context

Many years ago, when I was a much-impressed admirer of Arthur Rackham and his followers, I marveled at the wealth of contextual detail that they so artfully and carefully embroidered into the fiber of each picture. At the same time, I don't think I really understood very well the technique and effect at which I was looking.

The context (the background, if you will) can perform a number of different functions. Depending on the use of colors, textures, and light and dark, the effect can vary considerably. The subject may become part of a bewitching camouflage effect. The dynamism of the subject may be considerably reduced by the effect of the background even though its outer form and identity are still clear. Consequently, the decision simply to leave out the background was a bold one for me. In printmaking, this amounts to exchanging one set of problems for another: I found that I had to deal with background whether it was teeming with objects or consisted entirely of the eloquence of empty space. Many Oriental artists believe that it is the empty spaces that say the most.

ABOVE   Maple, *Acer macrophyllum*. Drawn from specimens collected in Bothe Park in the Napa Valley of California. Printed in 1973 on Hodamura paper; 175 copies. Size of block: 40.5 cm x 24.9 cm.

OPPOSITE   Persimmon, *Diospyros kaki*. Drawn at Sacramento, California. Printed in 1974 on Basingwerk paper; 150 copies. Size of block: 39.8 cm x 24.3 cm.

PAGES 44 AND 45   Pasqueflower, *Anemone patens*. Drawn at Missoula, Montana. Printed in 1972 on Hodamura paper; 125 copies. Size of block: 18 cm x 37.3 cm.

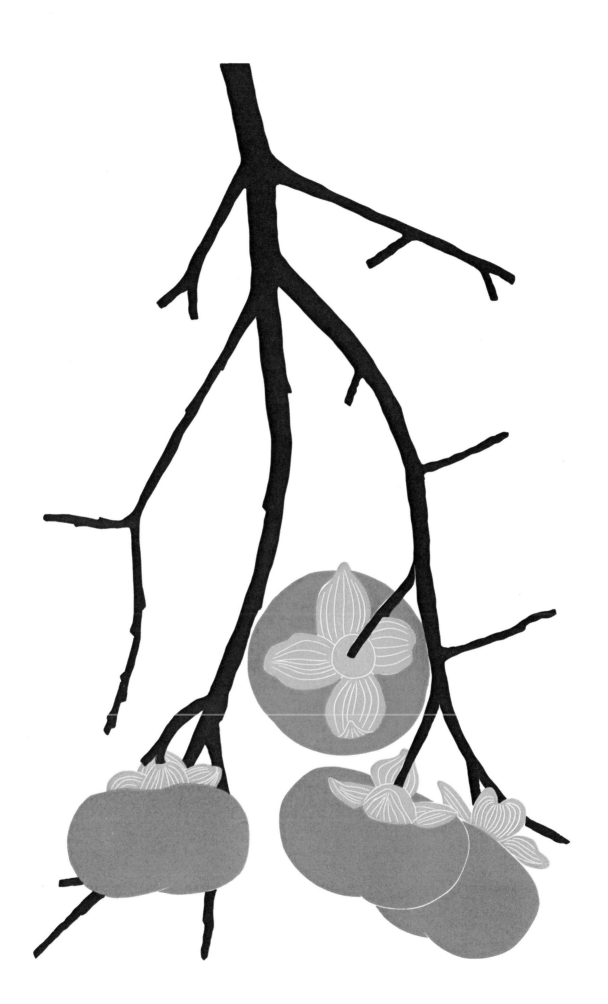

There are no doubt instances in which an artist would seek to create the illusion that his subject is afloat in space; more often, probably, it would be rewarding to be able to communicate a sense of lightness. In my work, I want to communicate to the viewer that this flower is alive and growing, that it is attached to parts of the plant that are not in the picture and, ultimately, to the earth.

In creating a picture, I try to transform the two-dimensional surface of the sheet of paper on which the image is to be printed into something seemingly three-dimensional—a surface that somehow has depth that one can actually look *into*. To create this illusion of three-dimensionality is one of the greatest achievements (tricks?) of art. It is a door, of sorts, to other dimensions, other kinds of existence. The whole near-far concept, the problems of perspective, the "where" of things—all of these depend on the artist's ability to put himself into the illusion of space he is trying to create. You must step in first before you can lead anyone else in after you. If somehow you do create the illusion of reality where before there was nothing, then you have a true sense of accomplishment. For the incurable problem solver this is deeply satisfying.

Recollection from Life

I tend to repeat some subjects—not usually the exact same specimen but frequently a similar one. For example, I have done more than forty California poppies in the past eighteen years. It has always been a popular subject, but more than that I felt a personal desire both to improve my treatment of it and to show some of the almost limitless forms and colors the plant seems to assume in the wild.

ABOVE Brodiaea, *Brodiaea laxa*. Drawn near the Dipsea Trail on Mt. Tamalpais, Marin County, California. Printed in 1973 on Vit Bloma paper; 165 copies. Size of block: 25.3 cm x 36.7 cm.

OPPOSITE Violet, *Viola papilionacea*. Drawing made in Barcelona in January of 1976. Printed in 1976 on Basingwerk paper; 220 copies. Size of block: 24.2 cm x 40 cm.

I have at times also returned to subjects that the public has found rather unexciting in order to answer my own personal needs. I simply felt I could do better and, as a matter of conscience, had to.

Sometimes when I have decided to do the exact same species a second time, I have gone out and looked for a very different specimen to draw, thinking I might encounter a more appealing plant or one with a more typical structure or form. Sometimes I have sought out a different specimen in order to get away from what I felt was a poor composition in the first print.

Once in a while I have simply returned to my original drawing and attempted to do a better job by redrawing it onto a new block. Very often this turns out to be a good way of making a better print. There have also been times when, after staring at one of my own prints hanging on the wall for several years, I have decided to make a new drawing with that *print* as the model. There is a left-to-right reversal problem with this, of course, but otherwise it is much the same as making a new drawing from an old original drawing.

Particularly with such plants as California poppies, wild oats, Japanese persimmons, and Douglas iris, I have simply turned on what powers of recall I have and have made a drawing with the aid of memory alone. After looking closely at many hundreds (more likely thousands) of specimens, it is possible to recollect (in tranquility) what the plant looks like. If I have already made many drawings of a particular species, I find that I can remember enough detail to make a good drawing. The main question in plant portraiture is not how to portray enough detail to make the plant identifiable to genus or to species, but rather how to know what to include and what to leave out to achieve both the right feeling and a believable image.

ABOVE   Bamboo, *Bambusa beecheyana*. Drawn at the Botanic Garden of the University of California, Los Angeles. Printed in 1973 on Basingwerk paper; 125 copies. Size of block: 40.3 cm x 24 cm.

OPPOSITE   Carnation, probably a horticultural form of *Dianthus caryophyllus*. Drawn in the studio in San Francisco. Printed in 1969 on Hodamura paper; 89 copies. Size of block: 40.2 cm x 10 cm.

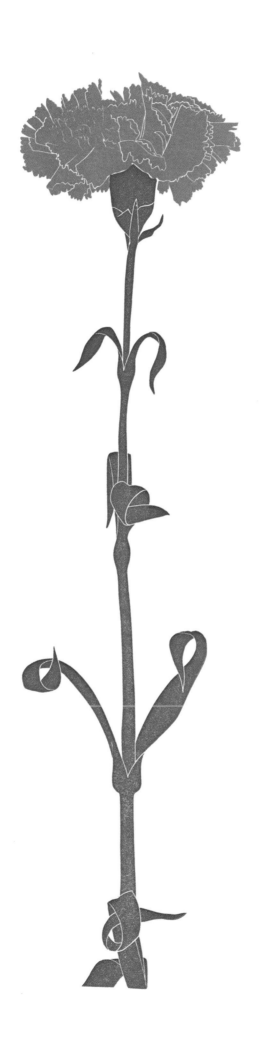

Essential Colors

After going through the selective process that relates to form, you are still faced with the job of determining which colors will be used, in what amounts, and where. When you look slowly and carefully at a plant, almost any plant, you will gradually become aware of the great number of colors, shades, and hues that are present in it. Like that of a woman who has applied her makeup carefully, the coloring of the plant, aside from the bright blossoms, is a very subtle matter, and the overall effect is achieved by very gradual color differences that one can see quite easily with a 10-power lens close up but that can never be seen from six or eight feet away.

Although some media are complex enough that they allow an artist to present nature with exceeding naturalness, the procedures of printmaking are such that normally one is obliged to select only a few of the most important colors with the hope of suggesting the overall feeling and demeanor of the plant. This enforced economy puts you on the spot, as it were. You must choose those colors that will create a strong *suggestion* of the original effect when printed adjacently (or overlapped). So the same problem exists in color selection for the print as in selecting how much of the plant (and how many details) to draw in establishing the form.

The difference in importance between form and color is very debatable—and probably irrelevant. Each is critical. Artists who use only black and white never have to deal with the problem. The rest of us must try to deal with it every day.

ABOVE   California Black Oak, *Quercus kelloggii*. Drawn from a specimen collected at Willits, California. Printed in 1975 on Basingwerk paper; 175 copies. Size of block: 25 cm x 39.5 cm.

OPPOSITE   Chrysanthemum, *Chrysanthemum morifolium*. Drawing made in London. Printed in 1976 on Basingwerk paper; 325 copies. Size of block: 40.5 cm x 25 cm.

Fortunately, plants are fairly consistent in their coloration. There are exceptions, of course, and one of the most obvious ones in my experience is the California poppy (*Eschscholzia californica*). Though the flowers are most commonly found in yellows and oranges, the total color range is from pure white to deep red.

Ordinarily the color differences between specimens are very slight. But an artist can use these differences to communicate moods and feelings.

The colors not only attract the birds and bees, they also attract us. But in our case, it is we who are, in a sense, fertilized.

Words to Describe Pictures?

A long time ago the Chinese made it clear to us how much better pictures could say it, but there are those artists who simply won't stop nurturing the delusion that they can say in words what they can't (or don't wish to) say in pictures. The artist can rarely know what those who look at his picture either bring to it or take from it. The picture remains intact, even in the face of mountains of criticism and intellectualizing. But the artist usually wants more than the praise and money his work may bring him. He very often wants somehow to make sure that whoever sees his work gets the "right" message from it.

When such artists as Goya, Daumier, Orozco, Siqueiros, Nast, and Young made strong political statements, their intent was so obvious that any booby could get the message. But when things get a little more subtle, as with a Manet, a Degas, a Lovis Corinth, or a Van Gogh, the bands of the political spectrum are not so important, and feelings become more emphasized. When this happens the artist has moved into areas more difficult to describe—perhaps more socially acceptable, but more difficult to write or speak about.

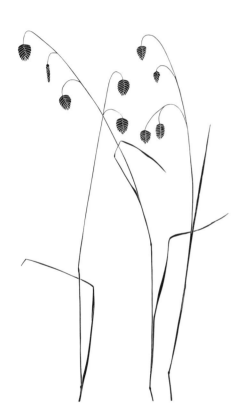

ABOVE Quaking Grass (or Rattlesnake Grass), *Briza maxima*. Collected at Pebble Beach, California, and drawn in the studio. Printed in 1975 on Basingwerk paper; 185 copies. Size of block: 39.5 cm x 23.5 cm.

OPPOSITE Sunflower, probably a horticultural form of *Helianthus annuus*. Drawn at Ukiah, California. Printed in 1976 on Basingwerk paper; 300 copies. Size of block: 41 cm x 25.5 cm.

Beyond this there is the inanimate subject of so many anthropomorphic meanderings—the flower, which exists light years away from politics, a symbol of human feeling but not directly part of it. So the presumption that someone can pick through his cerebral mare's nest and come up with tangent ideas, descriptions, and values sounds to me like a lot of far-fetched verbal wish-wash.

How neat it would be if a few dozen well-chosen words could convey the ethereal and evanescent essence of the most mysterious botanic feelings ever imagined. Unfortunately, it's a lot easier said than done. Very few real facts emerge from this quagmire: the only one I'm really sure of is that plants are beautiful *in themselves*—the why and the how are a great deal more difficult to explain.

Reproduction

No doubt one of the greatest joys (and worst terrors) for the artist is the notion of reproduction. He wants very much for his work to be widely seen (assuming he possesses a normal ego) and yet he is terrified of what the printers will do with his work.

Having had a number of years of experience as a printer of sorts, I can perhaps more easily understand the oft-repeated idea that there is a certain percentage of loss of fidelity at every stage of reproducing any work of graphic art. Each time there is a mechanical process interposed between the original and the final stage of the reproduction, a small amount of the detail and feeling is lost. The exact duplication of colors is never quite perfect. Most of the time the size of the original cannot be maintained. The surface texture of the reproduction is usually different from that of the original. Any reproduction is only an imitation of something you have worked hard to make, and when you see it in this lesser and distorted form many painful doubts may nag you.

ABOVE Anemone, probably a form of *Anemone coronaria*. Drawn at Brown's Hotel in London. Printed in 1975 on Basingwerk paper; 120 copies. Size of block: 39.5 cm x 23.5 cm.

OPPOSITE California Poppies and Brodiaea, *Eschscholzia californica* and *Brodiaea pulchella*. Drawn at Stanford University's Jasper Ridge Biological Reserve, San Mateo County, California. Printed in 1975 on Basingwerk paper; 150 copies. Size of block: 40.3 cm x 24.5 cm.

Botanical Portraiture

Some years ago in Pittsburgh during one of my earliest visits to the Hunt Institute for Botanical Documentation, the term *botanical portraiture* came to my attention for the first time. It was spoken by George H. M. Lawrence, then director of the Hunt Institute. The director of this unequalled repository of books, pictures, and other things botanical had impressed me with his wisdom and his learning as well as his suave and diplomatic style, so when he presented the term in question as a way of describing nonscientific botanical illustrations I immediately began to think about it.

Curators at the Hunt Institute have a handy rule about what they will include in their collections: the genus of the plants in any picture they acquire must be recognizable. I am sure that one could find pictures of plants in the Hunt collection whose genera might be arguable by botanists, but as far as those of us who earn our livelihoods by professions other than taxonomy are concerned, the rule is well adhered to.

The idea of combining the words *botanical* and *portraiture* is an intriguing one to me, for several reasons. It indicates that the picture is not scientific in its inclusion of (or respect for) minute detail. It also implies that the scale of reduction is not necessarily balanced or precise. The artist is relieved of the necessity of drawing such things as roots and dissections of the sexual parts, and he needn't even know the correct binomial. I don't think the term constitutes any sort of official stamp of approval on casualness or gross inaccuracy, but it does relieve the artist of the confinement of scientific drawing.

ABOVE Iris, *Iris douglasiana*. Drawn at the Sea Ranch, Sonoma County, California. Printed in 1975 on Vit Bloma paper; 170 copies. Size of block: 24.8 cm x 38 cm.

OPPOSITE Ebony Spleenwort, *Asplenium platyneuron*. Drawn in the studio from a specimen grown by Gunder Hefta. Printed in 1976 on Basingwerk paper; 450 copies. Size of block: 25 cm x 16.5 cm.

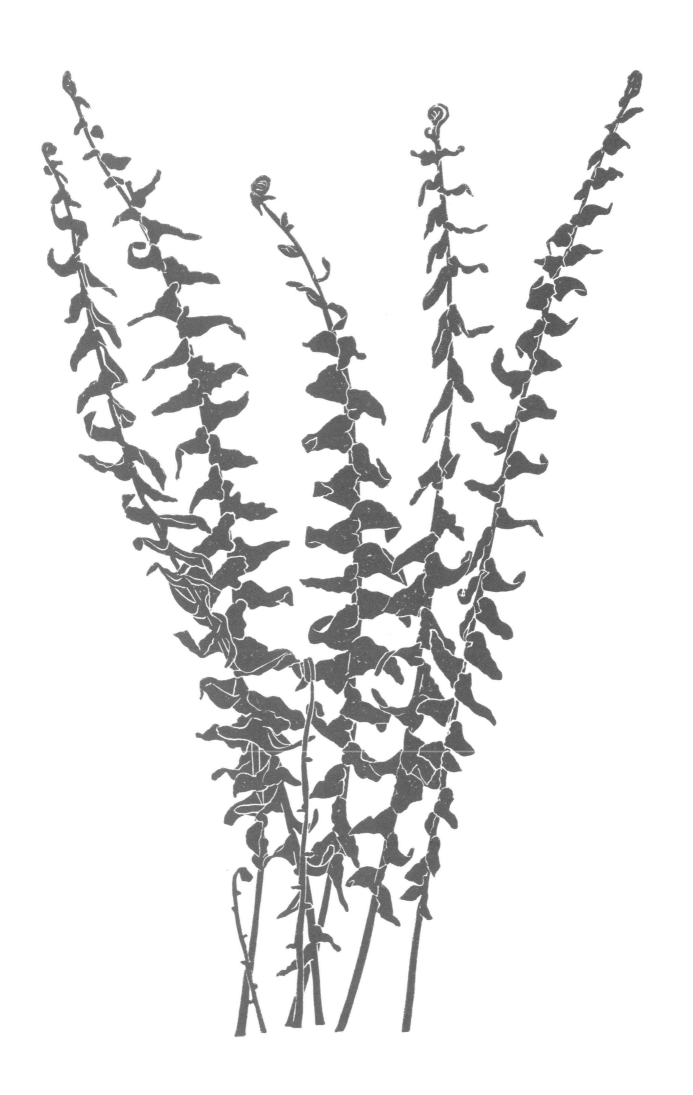

In the same way that the artist is given some leeway when a "likeness" of a person is wanted rather than an anatomical drawing, so the botanical portrait can display as much of the feeling of the artist as it does of the habit and personality of the plant. Another way to describe this difference might be to compare it with the difference between the idealized unblemished sculpture of the Greeks and the sculpture of the Romans, which included the warts, scars, rolls of fat, and all other distinguishing physical characteristics of the individual human being.

It's very easy for me to ignore the fact that some arrogant bug has chewed away parts of the leaf I am drawing—I simply draw the leaf as though it were perfect and whole. If a blossom is lopsided, I can straighten it up and make it look balanced. If a plant is a little on the runty side, I can deal with that. But if I were a scientific illustrator I would be operating under a somewhat different ethic; my pictures might be more like the plants really are and less like I would like them to be.

The Iron Hand Press

Unlike the common job press, which is powered by an electric motor, the hand press receives its activating force from the hand, arm, body muscle, and weight of the printer. The hand press is in essence a *tool* while the power press is a machine. The power press runs best at the optimum speed of the motor, whereas the hand press moves only at the speed the printer chooses. Over the years, books have been written about the virtues and peculiarities of each, but about one delicate matter no one who has operated both will argue: when a man or woman tends a power press, the press is the boss, and must be served constantly. When you are operating a hand press, however, every motion is decided by you. The hand press serves the printer.

Above   Redwood, *Sequoia sempervirens*. This twig of the coast redwood was drawn at the Sea Ranch, Sonoma County, California. Printed in 1976 on Basingwerk paper; 180 copies. Size of block: 12.5 cm x 11.7 cm.

Opposite   Violet, *Viola pedata*. Drawn at Gray Summit, Missouri. Printed in 1970 on Hodamura paper; 96 copies. Size of block: 20 cm x 15.5 cm.

Pages 60 and 61   Crocus, probably a horticultural form of *Crocus imperati*. Drawn in Amsterdam. Printed in 1973 on Hodamura paper; 160 copies. Size of block: 23.2 cm x 35.3 cm.

The hand press is capable of many adjustments and has inherent strength and durability. With reasonable care an iron hand press will last for centuries. Its parts do not wear out. The hand press has physical limits, according to its size, in how large a block (or how much type) it can print, but within those limits the excellence and subtlety of the product is entirely the result of the skill of the printer.

In the early stages of printing any block on the hand press, there are corrections and adjustments to be made. Usually, odd little bits and pieces need to be cut away from the block, and adjustments need to be made in the balance of the platen, as well as in the amount of pressure to be applied. Most early proofs are worthless and must be destroyed.

Once the basic adjustments are right, the matter of how much ink and the manner of its application must be dealt with. Sometimes it's better to have two or three thinner applications of ink than one heavier one. The eye must be trained. You must learn to see when the ink is getting too light; it diminishes somewhat with each impression. You have to be careful also not to over-ink. The work is spoiled by either too little or too much. It is also very important to develop your forearm muscles, in order to have precise control of the pressure of the ink roller on the block.

The iron hand press has not changed much since the time of Lincoln's childhood and the discovery of gold in California, and, despite what some people say, I don't think that people have changed much either. Running a hand press is essentially a matter of concentration and self-discipline. These qualities are no more remarkable now than they were a century and a quarter ago when the hand press was in its heyday. The only obvious difference is that there are a lot more people around and a lot fewer hand presses.

ABOVE Douglas-Fir, *Pseudotsuga menziesii*. Drawn at the Sea Ranch, Sonoma County, California, high on a coastal hill overlooking miles and miles of the Pacific Ocean. Printed in 1976 on Basingwerk paper; 180 copies. Size of block: 8 cm x 14 cm.

OPPOSITE Daffodils, *Narcissus* hybrid. Drawn in the studio in San Francisco. Printed in 1975 on Basingwerk paper; 140 copies. Size of block: 41 cm x 21 cm.

Pictures Replacing Words

The writer knows that the precise *meaning* of his words is an elusive matter. The artist knows (or will ultimately learn) that the invisible overburden each picture bears is really different for each viewer. Just as the most universally successful writers often use the simplest language, so the artists who are most admired and enjoyed choose subjects that are widely understood and easily perceived. Perhaps this explains why I have chosen plants. My reasoning has been that plants are a subject that most people can relate to with pleasure rather than trauma. I believe they offer a better way for me to express ideas about beauty, form, and mood than any other subject I could choose. There will never be a *Guernica* of botany. Even so, human feelings run very deeply in moments of tranquility. I would be satisfied to know that one of my pictures prompted the feelings that made possible one of these tranquil moments.

ABOVE   Madrone, *Arbutus menziesii*. Drawn on one of the slopes of Mt. Tamalpais, Marin County, California. Printed in 1973 on Basingwerk paper; 110 copies. Size of block: 24.3 cm x 39.6 cm.

OPPOSITE   Nasturtium, *Tropaeolum majus*. Drawn in Sausalito, California, in 1973. Printed in 1974 on Basingwerk paper; 125 copies. Size of block: 21.3 cm x 30.3 cm.

Composition by Mackenzie-Harris
in Monotype Original Old Caslon #337e.
Color separation by Color Tech Corporation.
Lithography by Fremont Litho, Inc.,
on 100-pound Tiara White Vicksburg Vellum
from Simpson Paper Company.
The book was edited by Gunder Hefta
and Linda Chaput. It was designed by William Tenney,
and its production was supervised
by Bruce Muncil and Jack Nye.